SEPT. 09

Test Results for Hardware Write Block Device:
T4 Forensic SCSI Bridge (FireWire Interface)

NCJ 228225

Kristina Rose

Acting Director, National Institute of Justice

This report was prepared for the National Institute of Justice, U.S. Department of Justice, by the Office of Law Enforcement Standards of the National Institute of Standards and Technology under Interagency Agreement 2003–IJ–R–029.

The National Institute of Justice is a component of the Office of Justice Programs, which also includes the Bureau of Justice Assistance, the Bureau of Justice Statistics, the Office of Juvenile Justice and Delinquency Prevention, and the Office for Victims of Crime.

Test Results for Hardware Write Block Device: T4 Forensic SCSI Bridge (FireWire Interface)

July 2009

**National Institute of
Standards and Technology**
U.S. Department of Commerce

Contents

Introduction

The Computer Forensics Tool Testing (CFTT) program is a joint project of the National Institute of Justice (NIJ), the research and development organization of the U.S. Department of Justice, and the National Institute of Standards and Technology's (NIST's) Office of Law Enforcement Standards and Information Technology Laboratory. CFTT is supported by other organizations, including the Federal Bureau of Investigation, the U.S. Department of Defense Cyber Crime Center, Internal Revenue Service Criminal Investigation's Electronic Crimes Program, and the U.S. Department of Homeland Security's Bureau of Immigration and Customs Enforcement, U.S. Customs and Border Protection, and U.S. Secret Service. The objective of the CFTT program is to provide measurable assurance to practitioners, researchers, and other applicable users that the tools used in computer forensics investigations provide accurate results. Accomplishing this requires the development of specifications and test methods for computer forensics tools and subsequent testing of specific tools against those specifications.

Test results provide the information necessary for developers to improve tools, users to make informed choices, and the legal community and others to understand the tools' capabilities. This approach to testing computer forensic tools is based on well-recognized methodologies for conformance and quality testing. The specifications and test methods are posted on the CFTT Web site (http://www.cftt.nist.gov/) for review and comment by the computer forensics community.

This document reports the results from testing the **T4 Forensic SCSI Bridge (FireWire Interface)** write blocker, against the *Hardware Write Blocker (HWB) Assertions and Test Plan Version 1.0* and *Hardware Write Blocker Device (HWB) Specification, Version 2.0*, available at the CFTT Web site (http://www.cftt.nist.gov/hardware_write_block.htm). This specification identifies the following top-level tool requirements:

- A hardware write block (HWB) device shall not transmit a command to a protected storage device that modifies the data on the storage device.

- An HWB device shall return the data requested by a read operation.

- An HWB device shall return without modification any access-significant information requested from the drive.

- Any error condition reported by the storage device to the HWB device shall be reported to the host.

Test results for other tools and devices can be found on NIJ's computer forensics tool testing Web page, http://www.ojp.usdoj.gov/nij/topics/technology/electronic-crime/cftt.htm.

Test Results for Hardware Write Block Devices

Device Tested:	T4 Forensic SCSI Bridge[1]
Model:	T4
Serial No:	000ECC010004D0B5
Firmware:	Jun 27 2007 09:40:43
Host to Blocker Interface:	FireWire
Blocker to Drive Interface:	SCSI
Supplier:	Tableau, LLC
Address:	N8 W22195 Johnson Drive, Suite 100
	Waukesha, WI 53186
	http://www.tableau.com/

1 Results Summary by Requirements

- **An HWB device shall not transmit a command to a protected storage device that modifies the data on the storage device.**
  For all test cases run, the device always blocked any commands that would have changed user or operating system data stored on a protected drive.

- **An HWB device shall return the data requested by a read operation.**
  For all test cases run, the device always allowed commands to read the protected drive.

- **An HWB device shall return without modification any access-significant information requested from the drive.**
  For all test cases run, the device always returned access-significant information from the protected drive without modification.

- **Any error condition reported by the storage device to the HWB device shall be reported to the host.**
  For all test cases run, the device always returned error codes from the protected drive without modification.

2 Test Case Selection

Since a protocol analyzer was available for the interface between the blocker and the protected drive, the following test cases were appropriate:

---

[1] Tableau produces this write block device for resale under various partner labels. See http://www.tableau.com for information on resellers.

- HWB–01
- HWB–03
- HWB–05
- HWB–06
- HWB–08
- HWB–09

For test case HWB–03, two variations were selected: file (attempt to use operating system commands to create and delete files and directories from a protected drive) and image (use an imaging tool to attempt to write to a protected drive).

3 Testing Environment

The tests were run in the NIST CFTT lab. This section describes the hardware (test computers and hard drives) available for testing.

3.1 Test Computers

Two test computers were used: **SamSpade** and **Max**.

SamSpade has the following configuration:

Intel® Desktop Motherboard FIC IC–VL67 (865G; S478; 800MHz)
BIOS Phoenix Award version v6.00PG
Intel® Pentium™ 4 CPU
Plextor DVDR PX–716A, ATAPI CD/DVD–ROM Drive
Western Digital Corporation WD800JB–00JJC0, 80 GB ATA disk drive
1.44 MB floppy drive
Three IEEE 1394 ports
Four USB ports

Max has the following configuration:

Intel Desktop Motherboard D865GB/D865PERC (with ATA–6 IDE on board controller)
BIOS Version BF86510A.86A.0053.P13
Adaptec SCSI BIOS V3.10.0
Intel® Pentium™ 4 CPU 3.4Ghz
2577972KB RAM
SONY DVD RW DRU–530A, ATAPI CD/DVD–ROM drive
1.44 MB floppy drive
Two slots for removable IDE hard disk drives
Two slots for removable SATA hard disk drives
Two slots for removable SCSI hard disk drives

3.2 Protocol Analyzer

A Data Transit bus protocol analyzer (Bus Doctor Rx) was used to monitor and record commands sent from the host to the write blocker. Two identical protocol analyzers were available for monitoring commands.

One of two Dell laptop computers (either **Chip** or **Dale**) was connected to each protocol analyzer to record commands observed by the protocol analyzer.

3.3 Hard Disk Drives

One SCSI interface device was used in testing:
- Drive label 25 is a SEAGATE ST373405LC with 143374741 sectors (73 GB).

```
Drive label: 25
Partition table Drive /dev/sda
143374741 total number of sectors
Non-IDE disk
Model (ST373405LC      ) serial # (3EK020ZB00002149H4DV)
 N    Start LBA Length     Start C/H/S End C/H/S   boot Partition type
 1 P 000000063 037752687 0000/001/01 1023/254/63 Boot 0C Fat32X
 2 X 037752750 067890690 1023/000/01 1023/254/63      0F extended
 3 S 000000063 063681597 1023/001/01 1023/254/63      07 NTFS
 4 x 063681660 004192965 1023/000/01 1023/254/63      05 extended
 5 S 000000063 004192902 1023/001/01 1023/254/63      06 Fat16
 6 x 067874625 000016065 1023/000/01 1023/254/63      05 extended
 7 S 000000063 000016002 1023/001/01 1023/254/63      04 Fat16
 8 S 000000000 000000000 0000/000/00 0000/000/00      00 empty entry
 9 P 000000000 000000000 0000/000/00 0000/000/00      00 empty entry
10 P 000000000 000000000 0000/000/00 0000/000/00      00 empty entry
```

3.4 Support Software

The software in the following table was used to send commands to the protected drive. One widely used imaging tool, IXimager, was used to generate disk activity (reads and writes) consistent with a realistic scenario of an accidental modification of an unprotected hard drive during a forensic examination. This does not imply an endorsement of the imaging tool.

Program	Description
sendSCSI	A tool to send SCSI commands wrapped in the USB or IEEE 1394 (FireWire) protocols to a drive.
FS–TST	Software from the FS–TST tools was used to generate errors from the hard drive by trying to read beyond the end of the drive. The FS–TST software was also used to setup the hard drives and print partition tables and drive size.
IXimager	An imaging tool (ILook IXimager version 2.0, February 2006) for test case 04-img.

4 Test Results

The main item of interest for interpreting the test results is determining the conformance of the device with the test assertions. Conformance with each assertion tested by a given test case is evaluated by examining the Blocker Input and Blocker Output boxes of the test report summary.

4.1 Test Results Report Key

A summary of the actual test results is presented in this report. The following table presents a description of each section of the test report summary.

Heading	Description
First Line	Test case ID; name, model, and interface of device tested.
Case Summary	Test case summary from *Hardware Write Blocker (HWB) Assertions and Test Plan Version 1.0*.
Assertions Tested	The test assertions applicable to the test case, selected from *Hardware Write Blocker (HWB) Assertions and Test Plan Version 1.0*.
Tester Name	Name or initials of person executing test procedure.
Test Date	Time and date that test was started and completed.
Test Configuration	Identification of the following: 1. Host computer for executing the test case. 2. Laptop attached to each protocol analyzer. 3. Protocol analyzers monitoring each interface. 4. Interface between host and blocker. 5. Interface between blocker and protected drive. 6. Execution environment for tool sending commands from the host.
Hard Drives Used	Description of the protected hard drive.
Blocker Input	A list of commands sent from the host to the blocker. For test case HWB–01, a list of each command code observed on the bus between the host computer and the blocker and a count of the number of times the command was observed is provided. For test cases HWB–03 and HWB–06, a list of each command sent and the number of times the command was sent. For test case HWB–05, a string of known data from a given location is provided for reference.

Heading	Description
Blocker Output	A list of commands observed by the protocol analyzer on the bus from the blocker to the protected drive.

For test case HWB–01, a list of each command code observed on the bus between the blocker and the protected drive and a count of the number of times the command was observed is provided. Also, a count of the number of unique commands sent (from the Blocker Input box) and a count of the number of unique commands observed on the bus between the blocker and the protected drive.

For test cases HWB–03 and HWB–06, a list of each command sent and the number of times the command was sent.

For test case HWB–05, a string read from a given location is provided for comparison to known data.

For test case HWB–08, the number of sectors determined for the protected drive and the partition table are provided.

For test case HWB–09, any error return obtained by trying to access a nonexistent sector of the drive is provided. |
| Results | Expected and actual results for each assertion tested. |
| Analysis | Whether or not the expected results were achieved. |

4.2 Test Details

4.2.1 HWB-01

Test Case HWB-01 Variation hwb-01 T4 Forensic SCSI Bridge FireWire	
Case Summary:	HWB-01 Identify commands blocked by the HWB.
Assertions Tested:	HWB-AM-01 The HWB shall not transmit any modifying category operation to the protected storage device.
HWB-AM-05 The action that a HWB device takes for any commands not assigned to the modifying, read or information categories is defined by the vendor.
Tester Name:	brl
Test Date:	run start Fri Oct 31 09:41:47 2008
run finish Fri Oct 31 10:09:51 2008
Test Configuration:	HOST: Max
HostToBlocker Monitor: Chip
HostToBlocker PA: AA00155
HostToBlocker Interface: FW400
BlockerToDrive Monitor: Dale
BlockerToDrive PA: AA00111
BlockerToDrive Interface: SCSI
Run Environment: Linux

Test Case HWB-01 Variation hwb-01 T4 Forensic SCSI Bridge FireWire		
Drives:	Protected drive: 25	
	25 is a SEAGATE ST373405LC with 143374741 sectors (73 GB)	
Blocker Input:	Commands Sent to Blocker	

Count	Commands
1	AC MANAGE
1	ASYNCHRONOUS CONNECTION
1	BLANK
1	CHANNEL USAGE
1	CHG DEFINITN
1	CLOS SESSION
1	COMPARE
1	CONNECT
1	CONNECT AV
1	CONNECTIONS
1	COPY
1	COPY/VERIFY
1	CREATE DESCRIPTOR
1	DIGITAL INPUT
1	DIGITAL OUTPUT
1	DISCONNECT
1	DISCONNECT AV
1	ERASE
1	ERASE(10)
1	FORMAT UNIT
1	GET CONFIG
1	GET EVNT/STS
1	GET PERFRMNC
1	INPUT PLUG SIGNAL FORMAT
1	INPUT SELECT
1	INQUIRY
1	LK/UNLK CACH
1	LOAD/UNLOAD
1	LOG SELECT
1	LOG SENSE
1	MECH STATUS
1	MEDIUM SCAN
2	MODE SELECT
2	MODE SENSE(10)
1	OBJECT NUMBER SELECT
1	OPEN DESCRIPTOR
1	OPEN INFO BLOCK
1	OUTPUT PLUG SIGNAL FORMAT
1	OUTPUT PRESET
1	PAUSE/RESUME
1	PERSISTENT RESERVE IN
1	PERSISTENT RESERVE OUT
1	PLAY AUD IDX
1	PLAY AUD MSF
2	PLAY AUDIO
1	PLAY CD
1	PLUG INFO
1	PLY TRK RLTV
1	PLY TRK RLTV(12)
1	PRE-FETCH
1	PREVENT/ALLOW MEDIUM REMOVAL
1	RD BUF CPCTY
1	RD GENERATN
1	RD MSTR CUE
1	RD STRUCTURE
1	RD SUB-CHNL
1	RD TOC/PMA
1	RD UPDATED BLK
1	READ BUFFER
1	READ BULK LIMITS
1	READ CAPACITY
1	READ CD

1	READ CD MSF	
2	READ DEFECT	
1	READ DESCRIPTOR	
1	READ ELEMENT STATUS	
1	READ FORMAT CAPACITY	
1	READ HEADER	
1	READ INFO BLOCK	
1	READ LONG	
1	READ REVERSE	
1	READ STATUS ATTACHED	
548	READ(10)	
1	READ(12)	
1	REASSIGN BLK	
1	RECEIVE DIAGNOSTIC RESULTS	
1	RECOVER BUFF DATA	
1	RELEASE(10)	
1	RELEASE(6)	
1	REPAIR RZONE	
4	REPORT KEY	
1	REPORT LUNS	
1	REQ VOL ADDR	
1	RESERVE	
1	RESERVE(10)	
1	RESERVE(6)	
34	RESERVED	
1	REWIND/REZERO	
1	SCAN	
1	SEARCH DESCRIPTOR	
1	SECURITY	
1	SEEK(10)	
1	SEEK(6)	
1	SEND CUE SHT	
1	SEND DIAGNOSTIC	
1	SEND EVENT	
6	SEND KEY	
1	SET CD SPEED	
1	SET LIMITS	
1	SET RD AHEAD	
1	SET STREAMNG	
1	SIGNAL SOURCE	
1	SND OPC INFO	
1	SND STRUCTUR	
1	SPACE	
2	SRCH DATA EQ	
1	SRCH DATA HI	
2	SRCH DATA LO	
1	SRCH DATAHI	
1	START/STOP	
1	STOP PLY/SCN	
1	SUBUNIT INFO	
1	SYNCH CACHE	
1	TEST UNIT READY	
1	UNIT INFO	
1	UPDATE BLOCK	
1	VENDOR-DEPENDENT	
1	VERIFY(10)	
1	VERIFY(12)	
1	VERIFY(6)	
1	WRITE BUFFER	
1	WRITE DESCRIPTOR	
1	WRITE FILEMARK	
1	WRITE INFO BLOCK	
1	WRITE LONG	
1	WRITE SAME	
2	WRITE(10)	
1	WRITE(12)	

	Test Case HWB-01 Variation hwb-01 T4 Forensic SCSI Bridge FireWire
	2 WRITE/VERIFY
	1 XDREAD(10)
	1 XDWRITE(10)
	1 XDWRITEREAD(10)
	1 XPWRITE(10)
	133 commands sent
Blocker Output:	Commands Allowed by Blocker

Commands Allowed by Blocker

Count	Commands
1	00h = TEST UNIT READY
2	03h = REQUEST SENSE
1	1Bh = START/STOP
1	1Eh = PREVENT/ALLOW MEDIUM REMOVAL
548	28h = READ(10)
1	2Fh = VERIFY(10)
1	35h = SYNCH CACHE

133 commands sent, 7 commands allowed

Results:	Assertion & Expected Result	Actual Result
	AM-01 Modifying commands blocked	Modifying commands blocked
	AM-05 HWB behavior recorded	HWB behavior recorded

Analysis:	Expected results achieved

4.2.2 HWB-03

Test Case HWB-03 Variation hwb-03-file T4 Forensic SCSI Bridge FireWire	
Case Summary:	HWB-03 Identify commands blocked by the HWB while attempting to modify a protected drive with forensic tools.
Assertions Tested:	HWB-AM-01 The HWB shall not transmit any modifying category operation to the protected storage device. HWB-AM-05 The action that a HWB device takes for any commands not assigned to the modifying, read or information categories is defined by the vendor.
Tester Name:	brl
Test Date:	run start Tue Nov 4 09:16:41 2008 run finish Tue Nov 4 08:54:36 2008
Test Configuration:	HOST: Max HostToBlocker Monitor: Chip HostToBlocker PA: AA00155 HostToBlocker Interface: FW400 BlockerToDrive Monitor: Dale BlockerToDrive PA: AA00111 BlockerToDrive Interface: SCSI Run Environment: Wxp
Drives:	Protected drive: 25 25 is a SEAGATE ST373405LC with 143374741 sectors (73 GB)
Blocker Input:	Commands Sent to Blocker | Count	Commands | |---|---| | 5 | MODE SENSE(6) | | 10 | READ CAPACITY | | 5791 | READ(10) | | 5 | TEST UNIT READY | 4 commands sent
Blocker Output:	Commands Allowed by Blocker | Count	Commands | |---|---| | 5 | 00h = TEST UNIT READY | | 5791 | 28h = READ(10) | 4 commands sent, 2 commands allowed
Results:	<table><tr><th>Assertion & Expected Result</th><th>Actual Result</th></tr><tr><td>AM-01 Modifying commands blocked</td><td>Modifying commands blocked</td></tr><tr><td>AM-05 HWB behavior recorded</td><td>HWB behavior recorded</td></tr></table>
Analysis:	Expected results achieved

4.2.3 HWB-03

Test Case HWB-03 Variation hwb-03-img T4 Forensic SCSI Bridge FireWire	
Case Summary:	HWB-03 Identify commands blocked by the HWB while attempting to modify a protected drive with forensic tools.
Assertions Tested:	HWB-AM-01 The HWB shall not transmit any modifying category operation to the protected storage device. HWB-AM-05 The action that a HWB device takes for any commands not assigned to the modifying, read or information categories is defined by the vendor.
Tester Name:	brl
Test Date:	run start Tue Nov 4 09:09:56 2008 run finish Tue Nov 4 09:55:09 2008
Test Configuration:	HOST: Max HostToBlocker Monitor: Chip HostToBlocker PA: AA00155 HostToBlocker Interface: FW400 BlockerToDrive Monitor: Dale BlockerToDrive PA: AA00111 BlockerToDrive Interface: SCSI Run Environment: IX
Drives:	Protected drive: 25 25 is a SEAGATE ST373405LC with 143374741 sectors (73 GB)
Blocker Input:	Commands Sent to Blocker Count Commands 2 INQUIRY 1 READ DEFECT 96 READ(10) 2529 WRITE(10) 4 commands sent
Blocker Output:	Commands Allowed by Blocker Count Commands 96 28h = READ(10) 4 commands sent, 1 commands allowed
Results:	<table><tr><th>Assertion & Expected Result</th><th>Actual Result</th></tr><tr><td>AM-01 Modifying commands blocked</td><td>Modifying commands blocked</td></tr><tr><td>AM-05 HWB behavior recorded</td><td>HWB behavior recorded</td></tr></table>
Analysis:	Expected results achieved

4.2.4 HWB-05

Test Case HWB-05 Variation hwb-05 T4 Forensic SCSI Bridge FireWire	
Case Summary:	HWB-05 Identify read commands allowed by the HWB.
Assertions Tested:	HWB-AM-02 If the host sends a read category operation to the HWB and no error is returned from the protected storage device to the HWB, then the data addressed by the original read operation is returned to the host.
Tester Name:	brl
Test Date:	run start Wed Apr 1 14:43:08 2009 run finish Wed Apr 1 14:54:26 2009
Test Configuration:	HOST: SamSpade HostToBlocker Monitor: Chip HostToBlocker PA: AA00111 HostToBlocker Interface: FW400 BlockerToDrive Monitor: Dale BlockerToDrive PA: AA00155 BlockerToDrive Interface: SCSI Run Environment: Linux
Drives:	Protected drive: 25 25 is a SEAGATE ST373405LC with 143374741 sectors (73 GB)
Blocker Input:	Commands Sent to Blocker Read sector 32767 for the string: 00002/010/08 000000032767
Blocker Output:	00002/010/08 000000032767
Results:	<table><tr><th>Assertion & Expected Result</th><th>Actual Result</th></tr><tr><td>AM-02 Read commands allowed</td><td>Read commands allowed</td></tr></table>
Analysis:	Expected results achieved

4.2.5 HWB-06

Test Case HWB-06 Variation hwb-06-img T4 Forensic SCSI Bridge FireWire	
Case Summary:	HWB-06 Identify read and information commands used by forensic tools and allowed by the HWB.
Assertions Tested:	HWB-AM-02 If the host sends a read category operation to the HWB and no error is returned from the protected storage device to the HWB, then the data addressed by the original read operation is returned to the host. HWB-AM-03 If the host sends an information category operation to the HWB and if there is no error on the protected storage device, then any returned access-significant information is returned to the host without modification. HWB-AM-05 The action that a HWB device takes for any commands not assigned to the modifying, read or information categories is defined by the vendor.
Tester Name:	brl
Test Date:	run start Tue Nov 4 10:12:16 2008 run finish Tue Nov 4 10:37:38 2008
Test Configuration:	HOST: Max HostToBlocker Monitor: Chip HostToBlocker PA: AA00155 HostToBlocker Interface: FW400 BlockerToDrive Monitor: Dale BlockerToDrive PA: AA00111 BlockerToDrive Interface: SCSI Run Environment: IX
Drives:	Protected drive: 25 25 is a SEAGATE ST373405LC with 143374741 sectors (73 GB)
Blocker Input:	Commands Sent to Blocker 	Count	Commands
2	INQUIRY
2	READ DEFECT
92	READ(10) | 3 commands sent
Blocker Output:	Commands Allowed by Blocker 	Count	Commands
141	28h = READ(10) | 3 commands sent, 1 commands allowed
Results:		Assertion & Expected Result	Actual Result
AM-02 Read commands allowed	Read commands allowed
AM-03 Access Significant Information unaltered	Access Significant Information unaltered
AM-05 HWB behavior recorded	HWB behavior recorded	
Analysis:	Expected results achieved

4.2.6 HWB-08

Test Case HWB-08 Variation hwb-08 T4 Forensic SCSI Bridge FireWire	
Case Summary:	HWB-08 Identify access significant information unmodified by the HWB.
Assertions Tested:	HWB-AM-03 If the host sends an information category operation to the HWB and if there is no error on the protected storage device, then any returned access-significant information is returned to the host without modification.
Tester Name:	brl
Test Date:	run start Fri Oct 31 14:41:01 2008 run finish Fri Oct 31 14:41:33 2008
Test Configuration:	HOST: Max HostToBlocker Monitor: none HostToBlocker PA: none HostToBlocker Interface: FW400 BlockerToDrive Monitor: none BlockerToDrive PA: none BlockerToDrive Interface: SCSI Run Environment: Linux
Drives:	Protected drive: 25 25 is a SEAGATE ST373405LC with 143374741 sectors (73 GB)
Blocker Output:	cmd: partab HWB-08 Max brl /dev/sde 25 -all 143374741 total number of sectors

Results:	Assertion & Expected Result	Actual Result
	AM-03 Access Significant Information unaltered	Access Significant Information unaltered

Analysis:	Expected results achieved

4.2.7 HWB-09

Test Case HWB-09 Variation hwb-09 T4 Forensic SCSI Bridge FireWire	
Case Summary:	HWB-09 Determine if an error on the protected drive is returned to the host.
Assertions Tested:	HWB-AM-04 If the host sends an operation to the HWB and if the operation results in an unresolved error on the protected storage device, then the HWB shall return an error status code to the host.
Tester Name:	brl
Test Date:	run start Fri Oct 31 14:27:36 2008 run finish Fri Oct 31 14:39:39 2008
Test Configuration:	HOST: Max HostToBlocker Monitor: none HostToBlocker PA: none HostToBlocker Interface: FW400 BlockerToDrive Monitor: none BlockerToDrive PA: none BlockerToDrive Interface: SCSI Run Environment: Linux
Drives:	Protected drive: 25 25 is a SEAGATE ST373405LC with 143374741 sectors (73 GB)
Blocker Output:	08923/254/63 (max cyl/hd values) 08924/255/63 (number of cyl/hd) 143374741 total number of sectors cmd: diskchg HWB-09 Max brl /dev/sde -read 1143374741 0 1 Disk addr lba 1143374741 C/H/S 71171/200/27 offset 0 Disk read error 0xFFFFFFFF at sector 71171/200/27

Results:	Assertion & Expected Result	Actual Result	
	AM-04 Error code returned	Error code returned	

Analysis:	Expected results achieved

About the National Institute of Justice

NIJ is the research, development, and evaluation agency of the U.S. Department of Justice. NIJ's mission is to advance scientific research, development, and evaluation to enhance the administration of justice and public safety. NIJ's principal authorities are derived from the Omnibus Crime Control and Safe Streets Act of 1968, as amended (see 42 U.S.C. §§ 3721–3723).

The NIJ Director is appointed by the President and confirmed by the Senate. The Director establishes the Institute's objectives, guided by the priorities of the Office of Justice Programs, the U.S. Department of Justice, and the needs of the field. The Institute actively solicits the views of criminal justice and other professionals and researchers to inform its search for the knowledge and tools to guide policy and practice.

Strategic Goals

NIJ has seven strategic goals grouped into three categories:

Creating relevant knowledge and tools

1. Partner with State and local practitioners and policymakers to identify social science research and technology needs.
2. Create scientific, relevant, and reliable knowledge—with a particular emphasis on terrorism, violent crime, drugs and crime, cost-effectiveness, and community-based efforts—to enhance the administration of justice and public safety.
3. Develop affordable and effective tools and technologies to enhance the administration of justice and public safety.

Dissemination

4. Disseminate relevant knowledge and information to practitioners and policymakers in an understandable, timely, and concise manner.
5. Act as an honest broker to identify the information, tools, and technologies that respond to the needs of stakeholders.

Agency management

6. Practice fairness and openness in the research and development process.
7. Ensure professionalism, excellence, accountability, cost-effectiveness, and integrity in the management and conduct of NIJ activities and programs.

Program Areas

In addressing these strategic challenges, the Institute is involved in the following program areas: crime control and prevention, including policing; drugs and crime; justice systems and offender behavior, including corrections; violence and victimization; communications and information technologies; critical incident response; investigative and forensic sciences, including DNA; less-than-lethal technologies; officer protection; education and training technologies; testing and standards; technology assistance to law enforcement and corrections agencies; field testing of promising programs; and international crime control.

In addition to sponsoring research and development and technology assistance, NIJ evaluates programs, policies, and technologies. NIJ communicates its research and evaluation findings through conferences and print and electronic media.

To find out more about the National Institute of Justice, please visit:

http://www.ojp.usdoj.gov/nij

or contact:

National Criminal Justice
 Reference Service
P.O. Box 6000
Rockville, MD 20849–6000
800–851–3420
http://www.ncjrs.gov